PREDATOR VS. PREDATOR

LION VS. HYENA

Lisa M. Bolt Simons

CAPSTONE PRESS
a capstone imprint

Published by Capstone Press, an imprint of Capstone
1710 Roe Crest Drive, North Mankato, Minnesota 56003
capstonepub.com

Library of Congress Cataloging-in-Publication Data
Names: Simons, Lisa M. B., 1969- author.
Title: Lion vs. hyena / by Lisa M. Bolt Simons.
Description: North Mankato, Minnesota : Capstone Press, [2022] | Series: Predator vs. predator | Includes bibliographical references and index. | Audience: Ages 8–11 | Audience: Grades 4–6 |
Identifiers: LCCN 2021012689 (print) | LCCN 2021012690 (ebook) | ISBN 9781663914101 (hardcover) | ISBN 9781663914200 (pdf) | ISBN 9781663914224 (kindle edition)
Subjects: LCSH: Lion—Juvenile literature. | Hyenas—Juvenile literature.
Classification: LCC QL737.C23 S559 2022 (print) | LCC QL737.C23 (ebook) | DDC 599.757—dc23
LC record available at https://lccn.loc.gov/2021012689
LC ebook record available at https://lccn.loc.gov/2021012690

Summary: The massive grassy African savanna is filled with the hunting activity of predators and their prey. The stamina of the hyena allows the animal to hunt its prey until it is too tired to fight. The lion's strength makes its attacks swift and severe. But which animal is the fiercest of the ecosystem? Compare and contrast each animal's features and skills before you declare the winning predator.

Image Credits
Alamy: Jim Varley Photography, 11; Minden Pictures: Denis-Huot / NPL, 8; Shutterstock: Albie Venter, 22, Alison Mees, 9, Alta Oosthuizen, 20, ANDRES MENA PHOTOS, 7, Andrey_Kuzmin, (scratches) design element, Apostrophe_S, Cover, middle 6, middle 10, middle 14, middle 18, Bonma Suriya, (texture) design element, Bridgena BArnard, 15, Catchlight Lens, top left 28, Chad Wright Photography, 5, COULANGES, 17, Haiko Meyer, 12, HaseHoch2, (lightning) design element, JacoBecker, 13, Jez Bennett, top 19, Kit Korzun, 4, Maggy Meyer, bottom 19, Mark Sheridan-Johnson, 26, MintImages, 27, Mogens Trolle, 16, PACO COMO, 23, Pfotenpaparazzi, top right 28, ReVelStockArt, (font) design element, SergeyBitos, (neon frame) design element, SerGRAY, design element, SteffenTravel, Cover, top 6, top 14, top 18, Stu Porter, 24

Editorial Credits
Editors: Gena Chester and Julie Gassman; Designer: Elyse White;
Media Researcher: Kelly Garvin; Production Specialist: Laura Manthe

Printed and bound in China. 5047

TABLE OF CONTENTS

Words in **bold** are in the glossary.

INTRO

Predators of the Savanna

The savanna is a **biome** formed by a huge area of **plain**. It has grasses, some trees and bushes, and rocks. It's perfect for predators like lions and hyenas to attack prey.

Hyenas and lions are **carnivores** and feast on many animals in the savanna. Most of the hunting for these two predators is done by females. Female lions are called lionesses.

A lioness on the hunt

A pair of spotted hyenas in South Africa

Because hyenas and lions eat other animals, they are predators. The animals they eat are called prey. Predators and prey depend on each other to survive in the savanna. The prey **evolves** as it tries to stay away from the predators. But the predators also evolve. They have speed, strength, **stealth**, and hunting skill to catch their prey and stay alive.

The Savanna

The warm savanna habitat is mostly flat and grassy. It rains a lot in the summer. Savannas cover many parts of Africa, Australia, South America, and India. Lions mostly live in Africa, but some also live in India. Most hyenas also live in savannas in Africa and India.

CHAPTER 1 Speed Limits

SPEED

STRENGTH

STAMINA & STEALTH

HUNTING ABILITY

LION SPEED

HYENA SPEED

Neither lions nor hyenas are the fastest animals on the planet. But they are certainly quick. They use their speed to catch their prey.

The lion can run up to 50 miles (80 kilometers) per hour. One of the lion's fastest prey is a small gazelle called a springbok. A springbok can run about 5 miles (8 km) per hour faster than a lion. But the lioness still rushes the prey and pounces!

Another fast prey is the wildebeest. It runs about the same speed as the lion. But the lioness's attack is quick!

A lioness chases down a wildebeest in Tanzania.

Telling a Tale with a Tail

Have you ever noticed a lion's tail? It's the only one of the big cats with a tuft, or a bunch of hair, at the end. The lion's tail is special because lions use it to communicate. Moving the tail can mean anything from commands to teasing.

Hyenas also have speed to catch their prey. Although the hyena isn't as quick as the lion, its speed is still effective.

A hyena can run more than 35 miles (56 km) per hour. The impala, a kind of antelope, is one of the hyena's prey. It can run about 55 miles (89 km) per hour. Even though the impala is faster, the hyena can chase it for long distances. The hyena chases the impala until the prey is exhausted.

Hyenas pursuing young gazelle

A hyena might lie down to rest after running long distances.

The hyena has a slower speed as well. It's called a trot. At 6 miles (10 km) per hour, the hyena can trot for long periods of time.

FACT

Hyenas are known for making a laughing sound. But they aren't laughing. The sounds are really a sign that hyenas are upset or stressed about something.

CHAPTER 2 Superior Strength

LION STRENGTH

HYENA STRENGTH

Lions and hyenas both have amazing strength to help their predator status. The hyena's jaws are one of the strongest of any mammal. A hyena may not be a huge animal, but its bite is powerful for its size.

When a hyena attacks an antelope, its jaws latch onto the animal. Then the hyena drags its prey to the ground and kills it. A hyena can take down an animal four times its size. They've even been known to attack rhinos.

Hyenas occasionally hunt rhinos.

Hyenas also have strong teeth. They can actually crush bones, even those in a huge African buffalo. Hyenas' stomachs are strong too. Their stomachs can digest bones and skin.

Like hyenas, lions are strong hunters. Once lions chase down their prey, they lunge at the prey's neck. Lionesses attack giraffes this way. They also do this with other prey, such as baboons. The hunters bite on a prey's neck to strangle the animal.

A lioness attacks a giraffe.

Lions have sharp fangs and powerful jaws.

Lions have teeth like scissors. They cut their prey into chunks. Then they swallow the chunks whole. As lions hunt and eat, they use their claws. When they're not being used, the claws **retract** into their paws.

CHAPTER 3 They've Got Game

SPEED

STRENGTH

STAMINA & STEALTH

HUNTING ABILITY

LION
STAMINA & STEALTH

HYENA
STAMINA & STEALTH

Lions and hyenas are both fast and strong. But they each have unique ways to use their hunting skills.

Hyenas are fast, but they can't outrun their fastest prey like wildebeest. Instead, they have **stamina**. The hyena can chase its prey for long distances. The wildebeest grows tired and can't run anymore. Then the hyena attacks.

A pack of hyenas chases off a lioness.

Lions are also more aggressive predators. If the two groups are competing over the same meat, a lion may hurt or kill a hyena. Lions and hyenas also fight over marked **territories**. On occasion, a lion has killed a hyena for reasons not known by scientists.

LION

VS.

HYENA

The lion is a fast-running hunter. The hyena has one of the strongest sets of jaws for its body size on the planet. The hyena has the better stamina. But the lion is more stealthy.

Both of the savanna predators are incredible. Which one would be named the top predator? Is it the lion? Or is it the hyena?

LION STATS

SHOULDER HEIGHT:
male lions: 4 feet (1.2 m)
female lions: 3 feet to 3 feet, 6 inches (0.9 to 1.1 m)

LENGTH: about 5 to 7 feet (1.5 to 2.1 m)

WEIGHT:
male lions: up to 500 pounds (230 kg)
female lions: up to 400 pounds (181 kg)

BIGGEST EVER RECORDED: 826 pounds (375 kg)

TYPES/SPECIES AROUND THE WORLD: one, with two subspecies: African and Asian

NUMBER OF TEETH: 30

PREY: African buffalo, antelope, baboons, elephants, fish, giraffes, hippos, warthogs, wildebeests, zebras

HYENA STATS

SHOULDER HEIGHT: about 29 inches (75 cm)

LENGTH: about 45 inches (115 cm)

WEIGHT: about 77 to 176 pounds (35 to 80 kg)

BIGGEST EVER RECORDED: unknown

TYPES/SPECIES AROUND THE WORLD: three: spotted, brown, striped

NUMBER OF TEETH: 32 to 34

PREY: birds, buffalo calves, fish, foxes, gazelles, impalas, jackals, ostrich eggs, porcupines, rabbits, warthogs, waterbucks, wildebeests, zebras

GLOSSARY

biome (BYE-ohm)—a community of plants and animals in a major region of the planet

carnivore (KAHR-nuh-vohr)—an animal that eats only meat

evolve (i-VAHLV)—when something develops over a long time with gradual changes

plain (PLAYN)—a large, flat area of land with few trees

retract (rih-TRAKT)—to slide back in

scavenger (SKAV-uhn-jer)—an animal that feeds on animals that are already dead

stamina (STAM-uh-nuh)—the energy and strength to keep doing something for a long time

stealth (STELTH)—the ability to move without being seen by others

territory (TER-uh-tor-ee)—the land on which an animal grazes or hunts for food and raises its young

READ MORE

Emminizer, Theresa. *Hungry Hyenas.* New York: PowerKids Press, 2021.

Klepeis, Alicia. *Lions on the Hunt.* Minneapolis: Lerner Publications, 2018.

Levy, Janey. *Hyenas Bite!* New York: Gareth Stevens Publishing, 2021.

INTERNET SITES

National Geographic Kids: Lion
kids.nationalgeographic.com/animals/mammals/lion/

National Geographic Kids: Spotted Hyena
kids.nationalgeographic.com/animals/mammals/spotted-hyena/

San Diego Zoo Kids: Spotted Hyena
kids.sandiegozoo.org/animals/spotted-hyena

INDEX

Hyenas can run for long distances.

The hyena takes advantage of its body during a hunt. It has a strong heart and lungs. It also has short and stocky legs. A long muzzle and wide nostrils help the hyena breathe and cool its body.

Hyenas also have stamina while hunting. They can walk 15 miles (24 km) to find prey.

A lioness hunts down a zebra.

A hunting lioness does not have stamina. Instead, it is a stealthy predator.

A lioness sees a zebra. Then she stalks the prey. Stalking means to walk slowly toward the prey or around it. The lioness watches the zebra. She moves closer. She waits 15 minutes to an hour. Then she speeds toward it. The lioness jumps and bites the zebra's neck.

Lions often hunt at night for better cover. During the day, the savanna grassland doesn't provide enough cover to hide behind. The wildebeest can see the lionesses easily. If there are bushes or rocky areas, the lions will use them for cover.

FACT

A person can hear a lion roar from 5 miles (8 km) away.

A lion roaring

CHAPTER 4 Top of the Animal Kingdom

SPEED

STRENGTH

STAMINA
& STEALTH

HUNTING ABILITY

LION
HUNTING ABILITY

HYENA
HUNTING ABILITY

A lioness is quick. A lioness is strong. A lioness is stealthy. What other abilities does a lioness have to make it a strong savanna hunter?

Lions can hunt with others or alone. Usually lions live and hunt in a group called a pride. The lions can circle herds of antelope. Or the pride can attack one large prey like the cape buffalo. The lions approach the cape buffalo from opposite directions.

Lionesses often work together to hunt prey.

When a lion hunts alone, it must be able to catch the prey by itself. One way the lion does this is by leaping. It can leap up to 36 feet (11 meters) to pounce on a blue wildebeest.

A lioness leaps to catch a cape buffalo.

Cooler temperatures make hunting at night more comfortable for lions.

A lion's incredible hunting ability doesn't stop there. Its eyes are more sensitive to light than human eyes. Its night vision is excellent. Lions can see six to eight times better than humans at night.

Lions can easily cross water while hunting. They even swim. Lions sometimes eat fish.

Lions hunt for about 2 to 3 hours a day. Like most predators, they may not be successful. But if they do catch a meal, they fill their bellies. An adult male can eat more than 75 pounds (34 kilograms) of meat! Then it doesn't have to eat for days or up to a week. It's time for the hunters to rest.

A hyena has speed. A hyena has strength. A hyena has stamina. What other abilities does a hyena have to make it a strong predator in the savanna?

Hyenas are able to hunt alone or in a group. Hyena groups are called clans or cackles. One hyena can hunt down a buffalo or zebra. A clan can even take down a young elephant. The size of a clan ranges from two hyenas to 30.

A hyena's jaws can crack through bones.

A pair of hyenas begin eating their prey.

Hyenas can and will eat anything. The acid in their stomachs can digest bones. If they eat hair, horns, or hooves, they throw these up later. Hyenas can eat diseased animal bodies and not get sick. They even eat poop!

FACT

Hyenas and lions have two, not three, cones in their eyes. This means they mostly see blues and greens. They don't see the color red.

Hyenas can cool themselves by wading in water.

What else makes hyenas amazing predators? They have good eyesight. It's actually best at night.

These savanna predators can swim across bodies of water to reach their prey. Some even sleep in shallow water. Fish can also be part of their diet.

Once hyenas catch their prey, it's time to eat. Even if they had to chase their prey for a long time, they don't need to rest before eating. They can eat between 30 to 40 pounds (12 to 18 kg) of meat for one meal. Then they rest.

Smart Predators

Hyenas are smart. They form relationships with others in their clan. They are able to solve problems together. The more scientists study hyenas, the more impressed the scientists are.

BUSTING A MYTH

When comparing lions and hyenas head-to-head, it's important to look at the facts instead of myths. In movies and books, hyenas are usually shown as **scavengers** and lions as hunters. But scientists have found the opposite to be true. Instead, lions are more likely to steal kills from hyenas. Hyenas hunt at least 50 percent of their meals.

A lioness fights with hyenas over meat.